GW01459565

By Constance Spry

THE ART OF ARRANGING FLOWERS

*With 25 reproductions in full color
and 24 in monochrome*

Constance Spry, who has had perhaps more influence on the art of flower arranging than anyone else living, both in this country and in England, presents the reader with a delightful collection of arrangements. She gives practical guidance on color combinations of flowers, how to make design arrangements in line and mass, and the selection of vases and suitable backgrounds. There are further notes on the care of flowers and other helpful tips for the lady of the household.

Having again distinguished herself—this time as official adviser on the flower arrangements for Queen Elizabeth's coronation—Mrs Spry now presents this enchanting book for every home where flowers are cherished. There are twenty-five lovely color subjects and as many in handsome monochrome. The foreword is written by the distinguished author Beverley Nichols.

THE ART OF ARRANGING FLOWERS

By the same author

FLOWER DECORATION
FLOWERS IN HOUSE AND GARDEN
A GARDEN NOTEBOOK
COME INTO THE GARDEN, COOK
SUMMER AND AUTUMN FLOWERS
WINTER AND SPRING FLOWERS

The Art of Arranging Flowers

CONSTANCE SPRY

With a foreword by
BEVERLEY NICHOLS

*Illustrated with
sixteen pages of color
and sixteen pages of black and white plates*

THE STUDIO PUBLICATIONS, INC
in association with
THOMAS Y. CROWELL COMPANY
NEW YORK

CONTENTS

		PAGE
Foreword by Beverley Nichols	. . . .	9
Introduction	. . . .	13
Flower decoration	. . . .	16
Choosing flowers	. . . .	20
Colours	. . . .	22
Line, proportion, and mass	. . . .	59
Vases	. . . .	63
Practical points	. . . .	66
Flower holders	. . . .	68

ILLUSTRATIONS

COLOUR

PAGE

An arrangement of foxgloves and columbines in a copper jug . . 27
Columbines in a silver bowl 28
Petunias and decorative kale 29
An arrangement of contrasting blues 29
A harvest festival group 30
Dahlias and coloured foliage 30
An arrangement of oak and horse-chestnut 31
Peonies and Philadelphus 32
Victoriana 32
Dahlias 33
Gypsophila in a wall basket 34
Rayonnante chrysanthemums and branches of elm . . . 34
Fruit and flowers 51
Honeysuckle on the wall 52
Elderberries and asters 52
A Flemish group 53
Hydrangeas 54
Lords and ladies 55
Sweet peas 55
A pale yellow arrangement 56
Zinnias and seed heads 56
Red roses 57
Massed summer flowers 57
Wallflowers in a wooden box 58
Michaelmas daisies 59

BLACK AND WHITE

Lilac, irises, and grey foliage 35
A black, white, and grey arrangement 36
Bluebells in an oval silver sugar bowl 36
Lilac 37
Wild parsnip and oats 38

A silver cup of early flowers 39
Nasturtiums 39
Anemones 40
Winter flowers 40
Blossom in a pottery jug 41
Hawthorn 42
Tulips and anemones 42
Poppies and willowherb 43
Primroses 44
Blossom 44
Madonna lilies 45
Pyrethrum 46
Sweet peas 47
Rhododendron and copper beech. 47
Yellow tulips and Solomon's seal 48
Single roses in a grey mug 48
A grey, blue, and mauve arrangement 49
Eschscholtzia 50
Dimorphotheca Aurantiaca 50

MAY I SAY . . .

I AM really the last person to write a foreword to this book, for I love it too much; and when you love a thing—or a person—too much, you are apt to go off into a flurry of italics, marks of exclamation, and breathless little sentences that end in three dots . . . There, you see, I have done it already.

However, Constance seemed to think it would be a good idea, and so do I if I can only hold myself in and keep up some sort of pretence of gentlemanly restraint. For it is always pleasant to pay tribute to an artist of such freshness and delicacy.

Constance Spry shares with another great woman—Ruth Draper —the distinction of adding a new phrase to the English language. We all know what it means, in the garden, to 'do a Ruth Draper,' for the simple reason that we have all done it or heard it done; we have all walked round our own or somebody else's borders, extolling the glories that have just faded or prophesying the glories that are about to be. And we have all made reference, during this shameless recital, to the name of the immortal Ruth.

Well, Constance is in the process of earning the same fragrant accolade. Time and again I have heard people say: 'It makes one want to do a Constance Spry.' The uninitiated, hearing so mysterious a phrase, might conclude that the speaker was on the point of whirling into some exotic dance, or absconding from his creditors, or taking to drink. The initiated—which means, of course, superior persons like you and me—know that it means something quite different.

It means standing before a bed of hydrangeas, when summer has fled, and seeing beauty in their pallid, parchment blossoms. It means suddenly stopping in a country lane, and noting for the first time a scarlet cadenza of berries, and fitting it, in one's mind's eye, into a pewter vase against a white wall. It means bouts with brambles, flirtations with ferns, and carnival with cabbages. Which

9

will be quite enough alliteration for the moment, considering that I am supposed to be 'holding myself in.'

Constance, in short, has the supreme gift—which is really the core of all art and all invention—of seeing things for the first time in a new way, and seeing them whole and seeing them isolated from convention.

Consider that chance phrase 'carnival with cabbages.' It means more than you might think, for one of her greatest services has been to break down the artificial barriers of prejudice which have been erected between the flower garden and the kitchen garden; she might indeed be described as the first floral artist who ever walked straight from the herbaceous border to the cabbage patch. Not *quite* the first, however; long before I knew Constance's work I myself had experimented by planting close clusters of blue pickling cabbages at the base of my delphiniums. This habit caused the gloomiest estimate of my morals to prevail in the neat little county of Huntingdonshire. There were some things they said that men—real men—do not do with cabbages, and this was evidently one of them.

Some of Constance's arrangements of the leaves of kale have the classic perfection of an early Florentine bronze; she achieves enchanting effects with purple-sprouting broccoli; and I should not be surprised if one day I found her doing something gay and delightful with a basket of Brussels sprouts.

Just as she has broken down the barriers which kept us from visiting the kitchen garden in our search for beauty, so she has opened the doors which kept us inside during the winter months, and shown us the delights that await us in the bleakest hedgerows on the darkest days . . . pale, spectral leaves, withered seedpods, berries of black and purple, bare branches. Of all her innovations I think the use of the withered hydrangea is perhaps the most significant; to-day it is almost a decorative commonplace; in the days of our grandparents it would not have been tolerated for a moment. Bells would have pealed, housemaids would have scurried, and the lovely, fragile, crinkly blossoms would have been pushed into the dustbin.

So it has been in every branch of flower decoration. Consider the flower vase. Twenty years ago it was *de rigueur* to jam every kind of flower, from polyanthus to *Lilium regale*, into a tall, thin glass vase, and hope for the best. These vases are still standard equipment in expensive nursing homes, no doubt because their effect is so depressing that they lower the morale of the patient, and thereby prolong his stay. But outside these establishments they are seldom encountered, and that is largely due to Constance. She has taught us that you should be as careful in choosing a vase for a flower as a dress for yourself, and she has widened the term 'vase' to include almost anything that is, in itself, beautiful, and capable of holding water.

If you feel inclined to challenge any of these statements, the many exquisite designs in this book are here to answer you. I do not propose to comment on them, or I shall burst into blank verse. They speak for themselves.

So I'll end by saying that I hope you love this book as much as I love it, and learn from it as much as I have learned from it. For Constance, bless her heart, is no airy theorist; she isn't like those monstrous females of the films who drift out into the garden in a floppy hat and come back twenty seconds later with a complete flower picture, which is dropped into a thousand-dollar vase with a single gesture. No siree! She may have gold in her heart but she has mud on her hands, and scratches too. And to her ardent labours a grey and troubled world is most deeply in debt.

BEVERLEY NICHOLS

INTRODUCTION

A LONG time ago my daily work took me into one of the poorest parts of a great city. The journey was dreary, long, and complicated, and I hated it. On many a spring and summer's day I used to take with me from the little garden of the cottage in which I lived a basket of flowers; according to the season the basket would be filled with pansies and pinks, roses and phlox, sweet rocket and wallflowers, primroses and daffodils, whatever might be in bloom. The extraordinary thing was that I don't think I ever got all the contents of the basket to their intended destination. One way and another they used to become scattered along my route. There was the bus conductor who had to have a pansy for his buttonhole because his grandmother grew them in the garden when he was a little boy, and there was the ticket collector who hadn't smelt mignonette for he didn't know how long, there were quite a few 'give us a flower, lady' urchins and one or two shyer ones who asked only with their eyes, and so, little by little, the contents of the basket dwindled. The journey, however, ceased to be a wearisome bore and became a comfortable, friendly affair enlivened at times with a bit of garden chat, for allotments had begun to come into their own; but more often with homely, youthful memories, personal memories of an earlier, more sunlit world, which added to day by day grew into sagas, nostalgic human sagas. That is one of the things flowers do for you, they break down barriers and make for friendliness, they crumble that wall of shyness that stands between so many English men and women, sometimes making their first reaction to a stranger hostile rather than friendly; you might perhaps call it the freemasonry of flowers.

These things, however, are not by any means all that flowers may do for you.

Many a man and woman during the bad and violent years that have fallen to us have found, through flowers and gardens, solace

13

from worry and strained nerves and escape from sordidness and anxiety. Allotments really took hold of us during the 1914–18 war and many learned for the first time the healing that comes through working in the soil and the joy and elation that is to be found in the miracle of growing a plant from a seed. Not that this little book is about the growing of flowers, alas! gardening is not possible for everyone: it is about arranging them. In the use of flowers for the decoration of her house the modern woman may find her artist's material. The creative urge is strong in us and among the strong emotions of the human heart is a love of beauty and a desire to create beauty. But it is not given to all of us to be artists in the generally accepted sense of the word. With flowers, however, with living plants as your medium, it is possible to create beauty, even to the degree of making a masterpiece. Quickly I want to say that this joyful experience is not limited to those who can grow or buy rare and expensive flowers, but is for everyone, school-child or student, town or countrywoman, for everyone who loves a beautiful thing and will take a little trouble.

Through the medium of one or two pictures I hope to show that decorative effects may be achieved with simple materials, that with a few leaves and grasses or weeds, as the Americans call wild flowers, there is a picture to be made. Other pictures showing more lavish use of materials may perhaps be of interest to those with a garden to pick from, these are offered as suggestions for the use of similar material, something for the reader to build upon. At no moment are they offered in the spirit of 'this is the way to do it.'

Indeed this is important to my mind, for it is all too easy to make rules and lay down laws, and by so doing to create barriers for beginners; to build up the idea that the arrangement of flowers is for the initiate. The truth is that an untrained child may create simple beauty; though I think it helps to be shown what may be called the 'tricks of the trade.' For the remainder individual imagination should be given free rein. In America, where flower arrangement competitions of a high and regimented order are very much the vogue, I have heard young women express the view that they would be afraid to arrange flowers which might be seen by members of the garden organization, because they would be sure to make mistakes and break rules. While I should be delighted to see such great enthusiasm about the arrangement of flowers in this country, I should be sorry if as a result we became rule-bound.

Some people may argue that to make a picture with anything so perishable as living flowers is disappointing because it is so soon gone, but the argument cuts both ways; your failures quickly fall into oblivion and your successes not only give infinite pleasure for the moment, but they remain a memory which long after the flowers are faded may 'flash upon that inward eye, which is the bliss of solitude.'

FLOWER DECORATION

FLOWERS in a room have a quality in common with the presence there of people or of a fire, they bring it to life, make it looked lived in. In addition to this they make a beautiful room more beautiful or a commonplace room interesting. I have even seen a Nissen hut, perhaps one of the least promising of backgrounds, decorated with simple flowers in such a way that people coming into it for the first time caught their breath in sudden surprise and pleasure. The eye leaps to a vase of flowers, leaps to something living and colourful. After the first instinctive glance of pleasure it is natural to look at the flowers more closely, more critically perhaps, and if on nearer view you find beauty of detail, good line, good colour blending, satisfying balance, your pleasure is so much the greater. I know that some people will argue that flowers are so beautiful in themselves that they can never look wrong, and when you take a perfect bloom in your hand you will feel moved to agree with them until you remember that when once you begin to use flowers to adorn a room you are using them as decorative material and not. as botanical specimens.

To use them well as decorative material you must think beyond the individual beauty of the bloom; among other things you must think in terms of good line and balance, good colour blending, and suitability of the whole to its background.

It is not enough to love flowers to arrange them well; it is very good first to love them, but beyond that it is necessary to work, to practise, to observe, to criticize, to be rarely, very rarely satisfied, to think and imagine and try again and again; and what infinite pleasure there is in it, and how much richer your everyday life becomes. It becomes richer because every hedgerow and field, every garden, little or big, every street barrow, flower-shop window, holds the material of your art or hobby or whatever name you choose to give it, and your imagination is fired and feeds on things that perhaps before you never even noticed.

16

I would like to talk a little about the materials of this hobby. Let us first take the point of view of those who live in and around big cities, for they are among the greatest of flower lovers. Cut off from the sights and sounds of the country one may long for flowers and leaves and growing things. I know that longing! In my younger days I used to assuage it in any simple way I could manage. I found that a celandine root would survive and open its flower buds in a saucer of water. I used to grow wheat or grass or canary seed in damp moss or earth in plant pots, saucers, or soup plates. The seed was scattered thickly over the surface, then the saucers or bowls were put away for a week or so in a dark cupboard until the growth was about a couple of inches high, when they were brought into the light and one could watch the delicate spears of clear green attain perfection; they lasted for weeks.

For those who will spend a little money on flowers they are to be found nowadays not only in flower shops but on street barrows and at the grocery stores. Although with a handful of flowers there is a picture to be made the enthusiast will not be content with these only but will augment them with less conventional material. A day in the country or a summer holiday may yield something that will last all the winter; it may be a bare branch or two to form a frame-work of a group of chrysanthemums or winter berries (the use of such a branch can be seen with chrysanthemums on page 34 and another with white tulips on the top of page 36). Perhaps the most valuable treasure trove to bring back from a holiday is a collection of shapely seed heads, especially the umbrella-shaped ones of such plants as wild carrot and parsnip (there is a picture of wild parsnip flowers on page 38). With these and a few berries you can make a gay autumn group. By themselves, if you have picked a few of contrasting shapes you can make something with the quality of an etching. And when their autumn use is fulfilled you can make a delicate frosted Christmas decoration. The point about the branches and seed heads is that they may be collected months before you want to use them and so are available to those who can escape only rarely into the country. For those who can get away often there are berries and grasses and wild flowers to add to such flowers as are to be found in the shops, and later I will write a paragraph on wild flowers for those who, like myself, have gathered them in hope, only to be cast down later with disappointment because they faded so soon.

From a bombed site I have seen a clever arrangement of grasses and seed heads and another of the fireweed or willowherb, that spiring, magenta flower that seems to spring up wherever fire has scorched the earth. This brilliant weed needs care if it is to last when picked and must be treated in the way described later on, but it is worth the trouble for it has, in addition to its own beauty, a quality which I find it hard to put into words. Added to a group of brilliant red flowers, it seems to resolve clashing colours into harmony. I cannot describe the delicate brilliance of a table decoration made last summer with this flower and mixed Shirley poppies. The poppies were in every shade of red, rose, and pink, and the cool translucent petals of the willowherb threw them into clean relief. It is one of the flowers that look particularly well against the light. There is an arrangement of these flowers on page 43.

And now in towns we have the blessing of allotments; strange that so good a thing should come out of war. They are, I know, chiefly for food production, but there is generally a corner reserved for a few flowers and there is good decorative material to be found among the most ordinary vegetables; carrots, parsnip, and kale leaves among others take on glorious winter colour, and even the most austere master of the allotment will spare these. Talking of vegetable leaves in decoration, may I say that if anyone tries to laugh at you or tease you out of using for decoration something which man has labelled for the kitchen just remind them that some orientals eat lilies as we would eat onions, and if that is not enough you might like to add that the beautiful decorative kales which we have begun to grow and appreciate have been used by the Japanese in flower arrangement for many decades, maybe, for all I know, for centuries.

Those who live in the country and have even a small garden with fields and lanes close by them have the wealth of the Indies to play with. But even there, with flowers at the door, it is a mistake to take only the more commonly used materials. A spray of crab-apples or of red currants will give contrasting form to a group of red flowers as will a branch of sloes or damsons to a blue one.

The five wine-red asters on page 52 were well set off by the elder-berries and the colour of the whole emphasized by the richly coloured beetroot leaves. The petunias at the top of page 29 gained a simple grandeur by the use of the open heart of decorative kale. It all sums up to this, when one is making a living picture it is good to choose freely from whatever provides the shape and colour one needs, without allowing oneself to be inhibited by convention.

There are, I think, two main ways in which a beginner may be helped, one easy, the other more difficult. The first or easy one may be described as the mechanics of arrangement and the care and preparation of material or—making flowers stay as you want them and helping them to last. The first being a matter of experience and of fact is fairly simple. The second is more difficult and is concerned with what may perhaps be called the aesthetic side. In this beginners may like to be started off, to be helped, by suggestions concerning colour, line, suitability, and general effect, but to go further than tentative suggestion is dangerous. Each individual must work to the pattern of his own ideas or there would be an end of originality; all that can be safely offered is a starting point.

I would like, in modern parlance, to debunk the idea that there are certain set rules of right and wrong for the arrangement of flowers. Such rules and opinions sometimes go to ridiculous lengths. Perhaps it is just plain obstinacy, but when I hear or read that certain colours should never be put together, or this class of flower be arranged with that, or am told that gypsophila should always accompany sweet peas, I feel the prison walls begin to close in, threatening the freedom of ideas; *freedom of ideas*, that is the important point. I think it is helpful to make suggestions and scatter ideas about, but it is mistake to be assertive about them, the danger is of creating barriers and limiting experiment.

CHOOSING FLOWERS

From shops. A progressive attitude in some flower shops makes it possible to buy half bunches or mixed bunches and this is a help. But where this is not possible, and when the budget is very restricted, one has to be clever about buying. A good plan is to buy leaves one week and flowers the next: sometimes the leaves outlast the flowers. I remember how good the flowers used to be in one apartment, owned by a woman who could not afford to buy many but who had a great sense of beauty. One good arrangement of leaves used to stand against a pale wall. Often it was of some form of evergreen, every leaf clean and shining and each branch well poised. It looked fine and handsome, and then somewhere, at some point of vantage, would be a small arrangement of flowers, perhaps of bright colour, or something sweet-scented. She bought wisely, took care of her flowers, and arranged them well.

From garden. In choosing flowers from a garden there is, of course, a much wider scope. Let us suppose you have in mind a red group, a picture in tones of red. You will walk about with this in mind and suddenly materials which you may never have thought of using come to your notice—a spray of loganberries, a branchlet of fruit, a red seed head, a brightly coloured kale leaf, some hawthorn berries or grasses. Among these will probably be just what you want to give contrast of shape and mass, and increased interest to the arrangement.

I would like to suggest too that you sometimes take flowers that are not among your favourites and work with them until something good is produced. This stretches the imagination, and the trying of conclusions with the widest possible range of material makes the whole subject more interesting. The picture on page 34 reminds me of one of my own foolish prejudices. For many a long year I left the annual gypsophila out of my range of materials. The word 'messy' springing to mind when I had thought about it, I left it alone. Perhaps the prejudice was built up by seeing it

20

used to fill up gaps in faulty arrangements of sweet peas or smudging up the dignity of roses or, rather tired looking, trailing among bouquets of carnations at weddings. Whatever the reason I let prejudice ride. Then one day, seeing a great sheaf of it in a local flower shop, and finding that for a few pence I could buy a good bunch of it, I decided it was time for me to practise what I preach and try to understand and use this flower *con amore*. It is truly delicate and exquisite in detail, with fine stems branching at good angles, and altogether of lace-like quality, but with certain small defects: the delicate stems break very easily, the open flowers curl up and die quickly, and the soft leaves wilt. If the sprays are left with curled flowers, broken stems, and wilting leaves, the effect is messy. I gave my bunches a long drink in a cool place, cut off every defective stem, flower, and soft leaf, and arranged the sprays in a wall basket, by themselves, endeavouring to emphasize the special quality of this flower in every way possible. The picture on page 34 was the result.

COLOURS

If you will agree that in making a flower arrangement you are creating a picture and not merely a botanical display, there are many points to consider. Covering them all is the important question of suitability to background and purpose: flowers, beautiful in themselves, can look wrong if they are out of keeping with their setting. Then follow points concerning colour, line, balance, contrast of mass, harmony between flower and vase, and various other considerations. It will be easiest to take each separately, so let us start with colour.

Choice of colour. Feeling for colour has its root in our whole make-up, in heredity even, and it is natural that these inborn instincts should guide us in our choice of colour, but it is a mistake to let them have complete sway. I feel inclined to say 'Try anything once!' If you don't—if, for instance, you say 'I don't like red or yellow or orange' or what have you, and leave these colours out of your flower adventures—you are losing something and in the end are so much the poorer. Do I sound as if I am always trying to persuade you to like something that you don't like? I don't mean it that way, I just don't want you to miss anything.

An interesting thing about colour in flowers is that it is possible to mix agreeably colours which in more opaque materials might be unpleasing. I think this is explained by the fact that translucence of petals and the consequent interplay of light affect the whole composition.

The effect of green on strong colour arrangements. Green, a lovely colour in itself, has the effect of cooling down other colours. If you are trying to achieve a strong note of colour, of red or orange perhaps, the incidence of green will detract from the brilliance of the result. Let us suppose that in a dull or cold room you choose to make a pool of golden marigolds, you will heighten the brilliance of the effect by the degree to which you dispense with their green leaves, though you may strengthen the effect of colour and warmth by the addition of coloured leaves. An example of this may be seen in the group of hydrangeas and vine leaves on page 54, in the picture of dahlias and leaves on page 33 and the one of zinnia and willowherb leaves on page 56.

SOME INDIVIDUAL COLOURS

Red. Red flowers have great popular appeal and are praised for their cheerfulness and gaiety. Striking and beautiful effects may be obtained by mixing together strongly contrasting shades of red and by adding perhaps red fruits and coloured leaves.

Suppose one wants to achieve a brilliant note of red in a room, perhaps on a cold day or in a north room, or perhaps to pick up a note of red in a picture, it is possible to get brilliance without hardness by putting together many shades and tones of red, rose, vermilion, crimson, magenta and, in the result, making a strong warm effect without harshness—something comparable in colour, as someone once said, with the blast of a trumpet.

Red flowers look well against walls of pale grey or egg-shell blue and, of course, against white. Clear magenta flowers like the wild willowherb are beautiful against pale yellow, as are many mauve, purple, and wine-coloured flowers.

Yellow. Pale yellow in particular seems to be spring's own colour. Pale primroses in their grey leaves and daffodils with their slender spears of green bring the early feasts of the flower year. The very words primroses and daffodils swing your mind into a spring song, 'primroses and daffodils and every meadow sweet,' and then you begin to think of apricot and gold azaleas and tall and stately tulips —moonlight comes into the room with a bowl of tulips called by that enchanting name—and then your mind runs on to yellow roses and lilies and a thousand and one flowers. On page 44 there is a picture of primroses and leaves in an old Sheffield plate cake-basket, and on page 58 yellow wallflowers in a wooden work-box. And on page 48 are cool, shapely yellow tulips with Solomon's seal, all of these flowers gathered, I would have you know, from an ordinary sort of garden. There is a group on page 53 in which yellow predominates, the flowers were picked and arranged on a hot summer's day; they looked refreshingly cool and had about them the touch

23

of an old Flemish flower painting. It is difficult not to grow enthusiastic about yellow flowers.

Blue, mauve, and purple. I think the beauty of blue reaches one of its highest points in a bluebell wood, perhaps after a shower of rain, so that the blue is misted over, and there are shafts of sunlight—for it is not a colour that lights up well, it tends to look grey. Blue, mauves, and blue-pinks or magenta look well together and light up well, and these colours together with purple look particularly well in silver or pewter. It may be in some cases they are helped by reflected light from the polished surface. I think this is particularly so with the hydrangeas on page 54 where the silver bowl of flowers is placed on a silver tray and one gets reflected light and colour. On page 58 there is an arrangement of Michaelmas daisies in a silver-plated pitcher, and on page 29 petunias in a pewter jug.

Green. It sometimes comes, perhaps after a feast of colour, that one longs for coolness, the coolness of green, and I confess I turn again and again to green groups and take great pleasure in arranging them. Leaves in themselves have architectural qualities which call for dignified and careful treatment. I remember that once or twice when we were doing the flowers for the British Pavilion at the World's Fair in New York we collected a group of enormous leaves, some of them the size of small tea-trays, and in many tones of green. In that great hall in a gigantic vase these seemed to me to have greater decorative value than the many wonderful flowers that were at our disposal during the period of the fair. For home decoration there are times and there are backgrounds that call for green. On page 31 there is a picture of the flowers of the common oak picked before the leaves unfolded, and one spray of horse-chestnut; soft lime-green oak flowers, deeper green chestnut leaves, and cream chestnut flowers, all this in a pale, shining brass bowl, make a quiet toned picture. At the top of page 55 is a silver sauce-boat of the green flowers of spring, lords and ladies or wild arums, green hellebores (most people love the white hellebores or Christmas roses), and a lovely, grey-green bell-shaped flower, called ornitho-galum, which is as easy to grow as a bluebell.

So often leaves are looked on as material of secondary importance, used for filling up the back of a vase or to hide gaps in a faulty arrangement. Of course they make a lovely foil for flowers (look

24

how they throw the dahlias into relief in the arrangement shown on page 30), but they also are lovely used by themselves. All through the winter I have had vases of ivy in my room, lovely, shining, elegant leaves, unmixed with flowers, delighting the eye. In America, where ivy is not so easy to grow as it is in England, it is more deeply appreciated and every flower shop sells pots of it. Laurel is another fine evergreen for decoration, used by itself grandly and not as a sort of stuffing for a few large chrysanthemums. But whatever evergreen you use, ivy, laurel, or any other, you should, before you arrange it, free every leaf of dust, using if necessary a pad of cotton wool dipped in warm water, and you should watch the vases to see when they need replenishing with water, for evergreens are thirsty things. For those who live in town green things may be more difficult, but the shops are getting better, and bunches of leaves and grasses can generally be bought and have the advantage of lasting a long time.

White. I would like to lay emphasis on the special quality of white flowers. I have heard them lightly dismissed as cold, funereal, and colourless. More often than not these opinions seem to have been accepted at second hand rather than to have come as a result of experiment. Rightly used and placed, white flowers are not cold, but a high light; not colourless but infinitely delicately shot through with tinted light and in some cases faintly reflective of the colour surrounding them. The picture on page 32 is of white peonies and philadelphus (sometimes called syringa) in a shell pink bowl. It does not seem to me to have the quality of coldness. 'Funereal' is the wrong word to apply to white flowers to-day, since modern fashion calls more and more for bright colour on sad occasions. If you feel prejudiced against white flowers, I would suggest that before shutting your mind to their appeal you might set white tulips or poppies against a delicate wall where strong light falls on them, or put a bowl of white flowers on a dark polished table, arranging them so that you get shadowy reflections in its shining surface; or you might fill an elegantly shaped bowl, like the one holding columbines on page 28, and ridding your mind of prejudice, look at white flowers again with new and considering eyes.

Mixed colours. Those who depend on shops or markets for their flowers can, generally speaking, more easily have flowers of one kind and colour for their vases, while for those of us who depend on gardens or allotments it is perhaps easier during most of the year to find a mixed bunch, either mixed flowers or mixed colours of one flower. Picking judiciously here and there one may find plenty of material and yet avoid leaving those gaps among the plants which are so apt to upset the master of the garden. In arranging these mixed colours or flowers there is just a point worth noting. One gets, I think, a better effect, at any rate a less fussy one, by introducing the colour in broad strokes rather than in spots. For example, in a bowl of mixed tulips or of sweet peas, flowers of one colour used in proximity give a cleaner effect than if they are dotted about in ones and twos. And in the case of mixed flowers, two or three flowers of one kind against two or three of another may give a better effect than if the flowers are put in in units.

Soft pinky-yellow foxgloves and deeper toned columbines with richly tinted columbine leaves are thrown into relief by the container—an old copper jug which has acquired a soft gleaming surface. See page 64. There are many wonderful colours among the cultivated foxgloves, which are easy to grow, excellent for cutting, and last a long time in water.

Cream coloured and pale yellow columbines in a silver bowl set on a silver salver. See page 26. To achieve a maximum decorative effect with a few flowers one may be helped by arranging them in this line—or this vase might form one of a pair.

Petunias and a large open heart of the purple decorative kale in a pewter jug.

An arrangement of contrasting blues.   Grape hyacinths, blue primroses, gentians, and forget-me-nots.

29

A group of fruit, autumn leaves, seed heads, berries, and flowers suitable for a harvest festival.

A silver birch log holding autumn foliage and mixed pompon dahlias. These neat dahlias last longer than many of the decorative types.

30

An arrangement of the lime-green flowers of the oak-tree
with a spray of horse-chestnut in a heavy brass bowl. One
often sees pussy willow and hazel catkins, but the flowers of
the larger trees are sometimes overlooked. These three branches
made quite an effective decoration. See pages 25 and 59.

31

Double white paeonies and white Philadelphus (syringa). Some of the heavier leaves of the syringa were removed, and the whole arranged in a shell-pink bowl.

Victorian china vases with old-fashioned flowers; Martagon lilies, pinks, sweet william, jasmine, poppies, columbine, old-fashioned roses, yellow marguerites, and blue butterfly delphiniums.

Decorative pompon and show dahlias with seed heads, a bare branch, and foliage in a copper urn. See pages 60 and 64. Dahlias are valuable cut flowers because they last so well—provided the water is not allowed to diminish below the stem line.

33

Annual gypsophila carefully cleaned of broken stems and all soft leaves arranged in a basket, with a lining to hold water, on a wall.

Rough branches of young elm form a framework for Rayonnante chrysanthemums; the leaves in the foreground are of purple-veined decorative kale.

34

Pale mauve lilac and iris with branches of grey
foliage from the sorbus or wayfarer tree in a pottery jar distempered to a pale bluish grey. The
leaves were picked in bud and allowed to unfurl
in the water. In this way they last a long time.

White tulips, grey leaves of begonia rex (from the greenhouse), and black ivy berries with a branch of cedar arranged in a silver dish.

Blue, pink, and white bluebells in a slender oval sugar bowl.   Good to set against the light of a window.

Various coloured lilacs in a silver jug.

The flowers of the wild parsnip and oats, arranged in a brightly polished copper jug. The seed heads of flowers of this shape are good for autumn and winter decorations. See page 17. Cow parsley and wild carrot are two other good umbrella-shaped flowers; they last well if preliminary care is taken.

Lilies of the valley, Iris sibirica, ladies' locket (Dicentra spectabilis), regal pelargonium, and lilac in a silver cup.

Nasturtiums and foliage in a shell. If these flowers are to look well they are best picked in bud and allowed to open in water.

Anemones and the riper leaves of the wild cow parsley; anemones often last for as much as two weeks and open their buds to perfection in water.

Christmas roses, ranunculi, snowdrops, and ivy with the early snake's-head iris (Iris tuberosa) in a plated entrée dish.

Double Japanese cherry in a tall white pottery jug. The flowers are wedged into place by the use of wire-netting. A narrow-necked jug is admirable also for sprays of the brilliant coloured 'japonica' (Cydonia japonica).

Hawthorn in a silver bowl. The old superstition about certain blossoms being unlucky might now be discarded. I believe it originated in fallacy and has robbed many people of pleasure.

Double pale pink tulips and richer toned anemones in a decorative china tureen.

Pink Shirley poppies, grasses, and slender willowherb—cool and refreshing colours for a summer's day. The flowers were picked overnight; the poppies, in bud, had the tips of their stems singed over a gas flame, the willowherb was stripped of its leaves. This treatment and a night in deep water before being arranged gave these summer flowers increased lasting power.

*An old-fashioned silver-plated cake basket is a good container
for pale yellow primroses and their leaves.*

*Soft pink blossom was here, in an opaque glass jar, set against a soft
rose curtain. The colour was bright without being harsh.*

Madonna lilies in a brass jar. A formal arrangement suitable for church decoration. See page 61. The lower leaves have been stripped off the lilies. When a number of flowers are arranged in a comparatively small container, care must be taken to replenish the water at least once and possibly twice a day.

The popular pyrethrum is not really easy to arrange in an interesting way. Here the rich red variety made the heart of a group, the outline and background of which is of branches of Prunus pissardii arranged in a copper coffee-pot. The pyrethrums were stripped of any foliage which proved too green for the colour scheme.

46

Sweet peas with trails of their own foliage in a shallow glass bowl on a round table.

Copper beech and pink rhododendron denuded of its green leaves, rich in colour and arranged in a copper fish-pan.

Pale yellow tulips (Niphetos) and Solomon's seal in a plated bowl—simple and cool.

Pale pink single roses in a grey mug. Single roses may not last as long as double ones, but all the buds will come out if preliminary care is taken.

Purple and mauve Michaelmas daisies, wine-coloured pompon dahlias, and a branch of blue-grey damsons in a royal blue china jug; for this arrangement many of the green leaves of the flowers were removed and the colour effect enriched by the addition of purple leaves of the decorative kale, a leaf much used by Japanese flower artists.

Deep golden eschscholtzias and rich brown grasses in a brass bowl.

The golden marguerite-like flowers of Dimorphotheca aurantiaca, an easily grown annual, are suitable for daytime decoration, for they go to sleep at night.

Flowers and fruit copied from a picture by George Lance (1802–64). A high vase is used for the grapes, vine leaves, and sprays of loganberries. Roses and passion flowers are arranged in a low bowl with blackberries and a small melon. The stems of the flowers and leaves are in water, but the fruit is protected from it by wire-netting. See page 62.

Trails of wild honeysuckle in a brass tinder-box used as a wall vase.

Elderberries stripped of their leaves, five wine-coloured asters, and coloured beetroot leaves arranged in a brass container.

52

A modern version of a Flemish flower group—yellow lilies, carnations, roses, corn, rich foliage, saxifrage, and vine leaves arranged in an urn. See pages 24 and 59. The rich effect is achieved with comparatively few flowers and helped by the use of dark-coloured leaves.

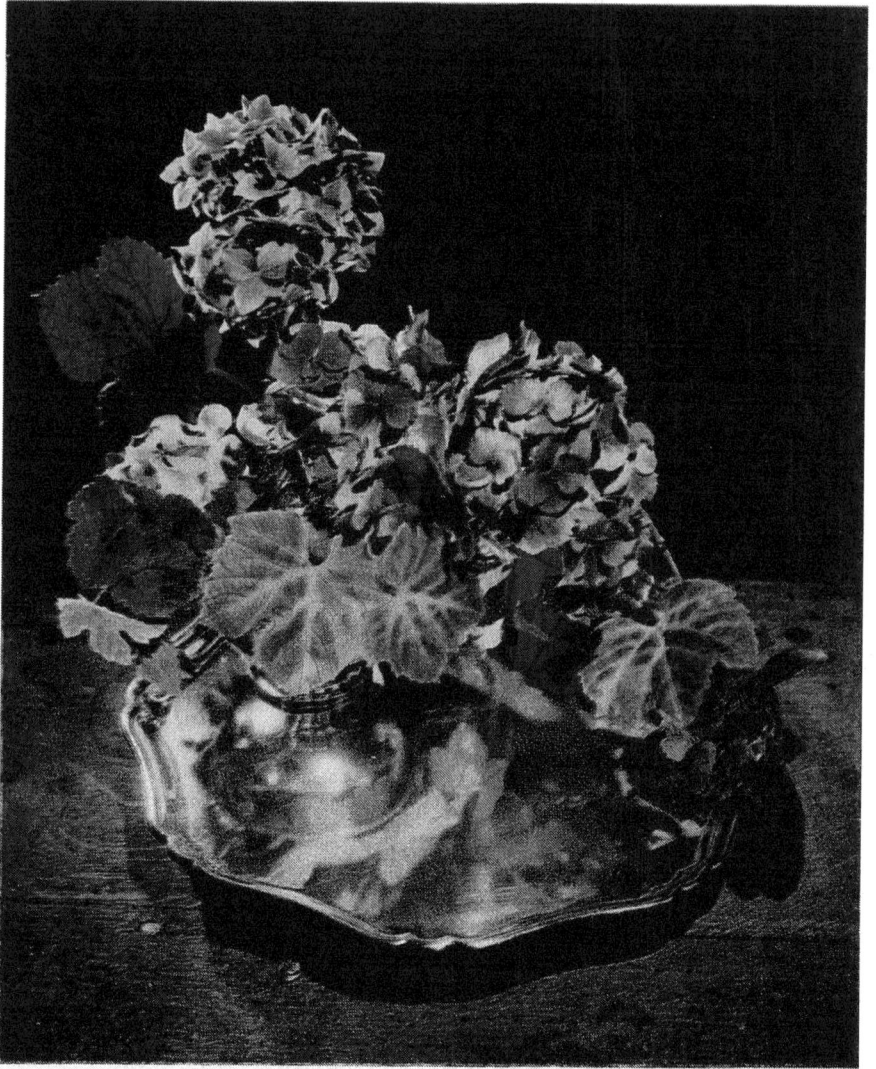

Hydrangeas picked when they have begun to dry a little on the plant take on wonderful metallic colours and last much longer in water than when picked earlier. Here they are shown with brilliantly coloured vine leaves in a silver bowl set and reflected on a silver salver. See pages 23 and 24.

A silver-plated sauce-boat filled with the leaves, flowers, and buds of wild arum (lords and ladies), a green tulip bud, green hellebores, and Ornithogalum nutans, a green-grey bell-shaped spring flower.

Sweet peas here are massed in solid groups of colour. An effective way of using them in a rich fashion.

Pale yellow marigolds among grey foliage. Velvety verbascum leaves and buds; striped anthericum leaves and grey artemisia in an old cake tin.

Zinnias and the leaves and seed heads of the wild willowherb or fireweed. Brilliant colours in a highly polished copper bowl.
(A warm arrangement for a dark room.)

Massed red roses in a delicate Georgian bread basket.   A glass pie-dish is used to hold water.

Massed summer flowers in a shallow copper cooking utensil.   Note that the flowers are put in small groups of one kind rather than in units.

57

Pale yellow wallflowers in a wooden work-box. Old wooden boxes provided with linings make good containers for certain types of flowers.

Mauve, purple, wine-coloured, and white Michaelmas daisies in a silver-plated pitcher. A good deal of green foliage has been removed.

58

LINE, PROPORTION, AND MASS

Line. The importance of line in flower arrangement is more easily illustrated by pictures than explained in words. An ugly or clumsy line may easily spoil a good colour group. It concerns not only the line of the flowers or their stems but also the outline of the arrangement as a whole. This is exemplified by the picture of oak on page 31, the nasturtiums on page 39, and the Flemish group on page 53.

There is a school of thought that sets so much store by outline that the flowers seem almost to be mangled in order that they may conform to the required shape. There is, I think, a half-way house by which flowers may follow a pleasing and suitable line, divorced from artificiality or from affectedness. To me the gentle curve of the columbines in the picture on page 28 is pleasant, though it would be possible to overdo this and to let it become an affectation. The outline of a group is generally influenced by the shape of vase or the size and shape of background. I find the simplest way to get the outline clear is to place in position two or three flowers to indicate the general shape I want to work to, putting in first perhaps the highest central flower and then the two sides, like this,

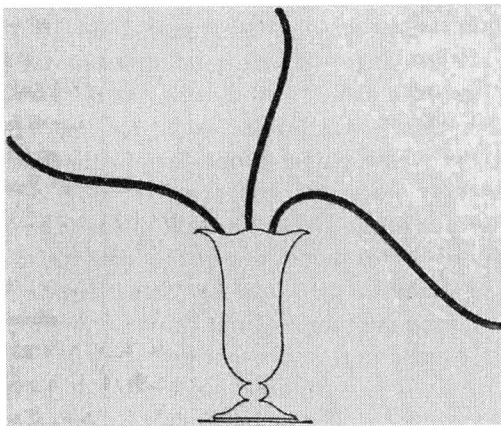

or for a cornucopia or a sauce-boat, putting the highest and lowest flowers, like this,

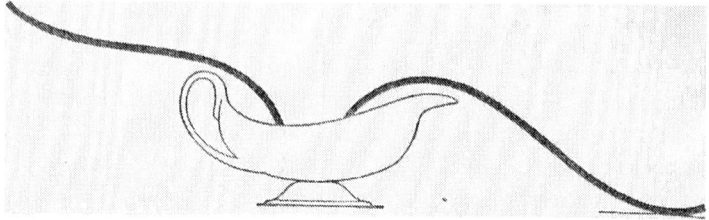

or for a centre-piece for a table, to fix first the total length like this,

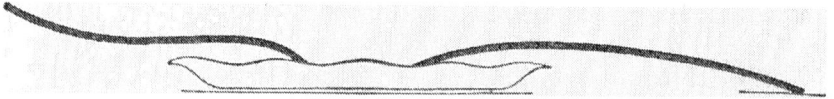

and then within such a framework it is easy to build up the whole.

The use of a shapely bare branch with certain flowers will often form a good outline. There is a rather formal example of this on page 36, where a cedar branchlet, black in colour, seemed suitable to accompany white tulips, black ivy berries, and grey begonia rex leaves. In the picture on page 33 you may see a bare brown branch on the left of the arrangement which went well with the tawny dahlias and autumn leaves, and on page 34 some rough twigs of young elm form a background to chrysanthemums. Yellow willow, red dogwood, young elm, larch—these and others in their leafless stage associate well with many flowers.

Proportion. If the flowers are too wide, too low, or too high for the vase, or the whole out of proportion to the background, the effect will not be good. Generally speaking, flowers a height of approximately one and a half times that of the vase may be taken as a rough guide. This should not be observed too literally, for on occasion flowers may be used in close massed formation. Another consideration is the character of the container. In a heavy piece of bronze or copper or brass a massive arrangement of flowers, leaves, and fruits will look suitable, while delicate china or glass can easily be overpowered. These points must be considered whenever it is intended that the vase shall be part of the picture: when harmony between flower and vase is important.

60

Background. The simpler the background the more clearly will the character of the flower arrangement be seen. When I can choose, I like best of all a whitewashed wall against which every quality of colour and line stands out. Plain green, for instance, or dark polished wood gives opportunity for strong contrast. I remember seeing white garden roses in a dark green room, and I think the lilies on page 45 look well against the dark wood.

Busy, ornamented backgrounds are a handicap, they interfere and confuse and so detract from the effect of the flowers. Where they exist and cannot be altered it is possible to help by choosing a permanent spot in the room for one's flowers, and hanging a piece of fabric or parchment as a background or by using a mirror. The Japanese sometimes use small twofold table-screens of parchment behind small groups of flowers to isolate them.

On the whole flowers are seen to their best advantage when set against an opaque background with the light shining on them. A mixed group of flowers when set against light may look confused. There are, of course, exceptions to this. Flowers with particularly translucent petals look charming against the light of a window. Single Shirley or Iceland poppies, bluebells, willowherb, are cases in point. The light shining through their petals brings the texture and colour of individual flowers into evidence. Where it is possible to light flowers well the effect is greatly heightened; it may simply be the light of a lamp shining down on them or, more elaborately, a spotlight. When such special lighting can be contrived the effect is really exciting.

The pleasure to be derived from a good group of flowers is heightened when it has relationship to some other object in the room, when, for instance, it echoes the tones of a picture or emphasizes the colour of curtains. Truly flowers have exquisite intrinsic beauty—a rose in a dungeon would shine out—but when they are being used in decoration they should be used with purpose, they should as it were have a reason.

I know a green room, so soft, quiet, and peaceful that a vase of bright carnations would destroy its quality. I know a bare attic-like room where one small vase of gay flowers makes it look furnished and decorated. I know also a modern chromium touched studio where only brilliant colour and clean-cut line have anything to say at all, where a tender arrangement of roses or pinks would be as unsuitable as a feather boa on a lift-girl's uniform.

And one more point: if you had to choose for the decoration of your room between one good picture or a dozen picture postcards you would, I think, choose the one good picture. So it is with flowers, one good arrangement is more effective than a number of little vases; this is one occasion when it is sensible to put all your eggs in one basket.

Mass. Contrast of colour is perhaps more generally considered than contrast of mass. On this point there is much for the lover of real flowers to learn from the pictures of the Old Dutch and Flemish painters. In their wonderful still-life masterpieces they grouped together massive fruit and flowers, delicate ears of corn, tendrils of vine, and fine grasses all so placed that the quality of each component shines out. To some extent we can copy this, putting perhaps a delicate frond of fern against a bold leaf, and so making a contrast of mass or using, as they so often did, fruit to give weight to the composition. On page 51 there is a group of fruit and flowers copied from a painting by George Lance (1802–64). This was arranged to show students at a summer holiday course that it was possible to get really near to the ideals of those old masters; a more practical adaptation of the idea for use to-day may be seen in the harvest thanksgiving group on page 30.

Detail. Since a living picture is being painted there should be no blot on the composition or none that one can see and avoid, so the last stem in place, it is well to stand back and satisfy yourself that all is well; that no part of the wire netting or other flower holder is showing, no stained water in view, no broken leaf or ugly line, no unintended gap; and, final precaution, it is well to see that the vase is filled to the top with water and that no leaf is so placed that it is acting as a siphon. Many a bit of damage has been done by neglecting this precaution.

VASES

Shape. It is worth while, I think, to say something about the shape of vases. If I could choose between an oval and a round vase, I should choose an oval because it is easier to arrange well. A boat-shaped vase is good on a chimney-piece or a window-sill. A chalice or goblet-shaped vase gives scope for a graceful, down-curving line. A wall vase has much to recommend it; it is economical of flowers, because every stem is silhouetted against the wall and made the most of, and there is no need to fill in the back and so confuse things. Fastened to the wall it will not overturn and can therefore carry spreading branches of some weight. There are good modern wall vases to be bought to-day, and sometimes one sees old-fashioned china hair tidies. I have seen copper measures cut in two and used in this way. On page 52 there is an old brass tinder-box filled with honeysuckle; this sort of vase is ideal for trailed climbing plants or sprays of drooping berries. I used a flat-backed bicycle basket on the somewhat unpromising walls of the Nissen hut referred to earlier. It was coloured to match the flowers, held jampots for water, and filled with a mass of the simplest flowers was a great success for a party. For schools, dance rooms, and hospitals there is something to be said for having flowers on the wall, out of harm's way.

Materials of vases. Glass, china, earthenware, metal, and basket-work—these are the chief materials concerned and each has its individual usefulness.

Glass. Clear glass vases are good to hold flowers with clean-cut stems. A rose stem showing through clear water, each thorn outlined with a lacework of minute bubbles, is delightful; but confused stems and stained water are an eyesore. The sweet peas on the round table on page 47 are in a shallow glass bowl, but the bowl is so low that only the rim can be seen, and that only in parts, so there is no danger of the stems showing.

China. Between blitzes and bombs and general wear and tear,

63

old bits of china become harder to find, and new fine china is still expensive, but if you have a good piece, an old bowl, a tureen or a vase, I would say keep it for special occasions, using it when it is exactly right for the flowers in hand. This way you may prolong its life and each time you use it you will see it with fresh eyes and enjoy it the more. The three little bits of Victoriana on page 32 are a delight to me during the summer when I can fill them with the same kinds of old-fashioned flowers with which the designer decorated them—moss roses and pinks, jasmine, sweet williams, and lilies. The shell in the picture on page 39 is a real shell, though one often sees them made in china, and a very good shape they are for any trailing flowers.

Earthenware. Jars, pitchers, honeypots in unglazed earthenware are inexpensive and often of good shape and they hold plenty of water. If their colour is not suited to the purpose they may be coloured. I find pale coloured distemper pleasant, but I do not like the effect of shiny oil paints.

There is an example of an arrangement in a distempered vase on page 35, and of single roses in a stone mug on page 48.

Metal. (Copper, brass, pewter, and tin.) One sees many metal flower containers used to-day. The shortage of good china may be one reason for this, but I think it is more likely that flower arrangers appreciate the good points of copper and brass and pewter. Jugs, urns, bowls, and preserving pans of good shape make most satisfactory vases. Heavy in weight and generally holding plenty of water, they form a secure base for spreading arrangements and for the larger flowers. The gleaming surface, high lights and reflections, all add to the general effect. On page 56 may be seen zinnias in a copper bowl, on page 50 eschscholtzias in a brass one; the pinky-yellow foxgloves and columbines on page 27 were tried first in a pottery vase and were disappointing, but when I transferred them to the old copper jug they came to life. The copper urn on page 33 is a satisfactory container for the massive group of autumn colour. The pewter jug on page 29 is good for many things—yellow roses, grey leaves, and Michaelmas daisies among others; and on page 56 are yellow and grey flowers arranged in an old cake tin which has been many times in a hot oven and has taken on a fine surface.

64

Silver. The use of silver brings us back to convention for it held pride of place for many a long year. I should find it hard to arrange flowers as I like to have them in the trumpet vases and rose bowls once so popular. But among silver and silver plate are to be found delightful shapes to hold flowers. English silver is pre-eminent in beauty and fortunately many good shapes and patterns have been copied in plate.

A rectangular or oval entrée dish is excellent for a low table arrangement; a pair of gravy boats give opportunity for variation from the usual circular or oval centre-piece. The Georgian bread-basket at the top of page 57 is a lovely container for massed red roses and, being low, it allows one to use tall candelabra, so that soft light can shine down illuminating the flowers.

Baskets and wooden containers. Simple baskets and trugs make excellent containers for mixed garden and wild flowers and for homely things like wallflowers and polyanthus, garden roses and marguerites. They need a tin lining or a baking- or bread-tin to hold water. On page 58 is a picture of yellow wallflowers in an old wooden work-box. These old boxes and tea-caddies can thus be put to practical use; they need, of course, to be provided with a lining. A log of silver birch hollowed out in the centre and supplied with a tin lining holds the dahlias and foliage shown on page 30.

PRACTICAL POINTS

The following points may help a beginner:

1. Whether flowers are picked from the garden or bought from a shop it is a mistake to put them straight away in their vases, they will last better if they can first have a long drink in a cool place. In some cases they may with advantage be left in deep water all night, before being arranged.

2. Flowers with woody or hard stems should have the tips of the stems hammered or split for an inch or two; this enables them more easily to absorb water. This applies to roses, chrysanthemums, lilac, and the blossoms of many flowering trees and shrubs.

3. If flowers show signs of wilting after a long journey, or because they have been too long out of water, the stem may be put into hot water. This is particularly the case with hardwood subjects such as roses, lilac, chrysanthemums, and fruit blossom.

4. Certain flowers and blossoms which carry a large amount of foliage in proportion to flower will last better if some of the foliage is removed: lilac and syringa (*Philadelphus*) are examples. In the case of garden lilac it is sometimes advisable to remove all leaves from the flowering stems, adding a separate spray of foliage when required for effect. It would seem that the heavily leafed stems are unable to absorb enough water to keep both flowers and leaves alive. Leaves which come below the surface of water and are not required should be removed. They decay quickly and make the water unpleasant.

5. Flowers dislike draught and the dry air which comes from radiators and gas and electric fires.

6. Vases should be filled to the brim when all the flowers are in place. They should be filled up again a few hours later and kept filled. It is not necessary to empty and refill. This disarranges the flowers and is apt to bruise them. Once they are in place they should be disturbed as little as may be. Where there are many flowers in a vase especial care should be taken about filling up for the vase very quickly becomes dry.

66

7. If flowers such as tulips or rosebuds show signs of wilting they may be lifted out of the vase, have their stems re-cut, be rolled up in newspaper to keep them straight, and be plunged to the neck in water in a dark cool place until they have revived. Poppies and bluebells should have the tips of their stems dipped into boiling water for a moment, but only the tips. In the case of poppies the tips may be singed instead over a gas ring. Hellebores last better if allowed to 'swim' in water for a few hours before being arranged.

All the precautions mentioned apply particularly to wild flowers, which should be wrapped up for the journey home and put immediately into deep warm water to revive.

FLOWER HOLDERS

IT is essential that there shall be some mechanical means of making a flower stay in the position required, and to be able to have a flower at any angle you wish. Many flower holders have the fault of making flowers look like hatpins in a pincushion.

One of the simplest and most efficient holders is made with wire-netting. A large mesh and thin wire are most suitable. This should be crumpled up into a ball of the size and shape suitable to the vase. For a tall opaque vase it is sufficient to push the wire into it; for a shallow bowl it may be necessary to tie the wire in with string, like a parcel. After the flowers are arranged the string may be cut away. For a glass vase the wire may be kept near the top and hooked over the rim; it is possible then to hide it with an overhanging leaf or flower.

A useful type of flower holder for small arrangements has now come back on the market. I believe it originally came to us from Japan. It consists of a heavy metal base closely covered with many sharp needle-like spikes which penetrate the base of the flower stem and hold it firmly in place. It has its greatest sphere of usefulness for flowers arranged in shallow bowls or chalices or even in still shallower meat-dishes or soup plates. It is excellent for medium-sized flowers, such as daffodils, early tulips, and a few roses. I have used the larger sizes to hold as much as two or three arums and a leaf or two provided there was not too much leverage from length of stem. With this as with all other holders care must be taken to hide it from sight, for it is utilitarian and not decorative.

Finally, may I add a word generally about arrangement. First decide the height and width and work to these, but be chary of shortening the stems of the flowers. Wherever it is possible to achieve the desired variations in height by pushing the stem further into the vase either vertically or horizontally, do so rather than gain your end by chopping off a piece of the stem. There are several reasons for this: in the first place the stem seems to take a more natural flowing line when it is not cut off; secondly, it is more likely to remain permanently below the water line, and thirdly, you cannot put the stem on again as you might wish were possible, if by chance you changed your mind.   I would like also to say : Keep an observant eye and an open mind, never stunt in order to be original.   And may I add, if I do not sound dictatorial, that while it is difficult not to get into a habit of arranging flowers in a set and limited number of ways this may be avoided by keeping an ever critical eye on all one's own efforts.

DATE DUE

MAR 2 5 1992			
GAYLORD			

Printed in Great Britain

CPSIA information can be obtained
at www.ICGtesting.com
Printed in the USA
LVHW082013090522
718307LV00005B/209

9 781014 252906